Title: Renewable Energy - A Buyers Guide

Author: Marc Asker

Published by PD Youle-Grayling

© Marc Asker 2014

All rights reserved. No part of this publication may be reproduced or transmitted in any form or by any means, electronic or mechanical, including photocopy, recording, or any information storage and retrieval system, without written permission from the author.

Disclaimer

The author does not accept responsibility, in any manner whatsoever, for any error, or omission, nor any loss, damage, injury, adverse outcome, or liability of any kind incurred as a result of the use of any of the information contained in this book, or reliance upon it.

Acknowledgements

My sincere thanks to the following companies for providing information, pictures and their support in producing this book.

Hyrax Solar Power Company, Grimsby, DN31 2TG

www.hyraxsolar.com

Lailey & Coates, Slough, SL1 4NH

www.laileyandcoates.com

Go Geothermal, Newton Aycliffe, DL5 6SP

www.gogeothermal.co.uk

Brew & Corkill, Douglas, Isle of Man, IM2 4BL

www.brewcorkill.co.im

SCS, Baldrine, Isle of Man, IM4 6AJ

www.scs.co.im

Contents

What is Renewable Energy

Why use Renewable Energy

Solar Thermal systems

Solar Photovoltaic

Wind Turbines

Micro Hydro

Micro CHP (Combined Heat and Power)

Mixing Technologies

PVT (Solar Photovoltaic / Solar Thermal)

Space Heating

Space Heating Emitters: Standard Radiators

Space Heating Emitters: Smart Rads

Space Heating Emitters: Underfloor Heating (UFH)

Heating Controls

Heating Controls: Thermostatic Radiator Valves (TRV's)

Heating Controls: Programmable TRV's

Hysteresis, Optimised heating controls and the Internet

A word about Condensing Boilers

Thermal stores

Heat Pumps

Air Source Heat Pumps

Ground Source Heat Pumps

Hybrid Heat Pumps

Biomass - more than just wood

Voltage Optimisation

Summary

Preface

Spiraling energy costs, climate change, security of energy supply - things that we weren't that concerned about just a few years ago but which now constantly make the news.

If you're planning to spend on Renewables (or just want to know more) this guide provides an introduction to the various products and technologies available for the technically (and not so technically) minded.

Written to provide an understanding of various renewable energy technologies and energy saving methods - the issues, economics, how they are applied, what they can and of equal importance, what they can't do.

This book has been written with UK standards, tariffs, planning and so on in mind.

However, as far as the technologies go these are the same worldwide - physics is physics - so wherever you reside I hope you will find this guide useful.

What is Renewable Energy?

One of the best descriptions I've seen is *"Energy flows which are replenished at the same rate as they are used"* which is from the BPEC* instructors training presentations.

For example, if we use solar power we won't deplete the suns energy by doing so - the energy we use will be replenished.

Interestingly most renewable energy comes from solar energy either directly - solar water heating, solar electricity, ground source heat pumps etc., or indirectly from hydro, wind or biomass.

Fortunately we can exploit the power of the sun and not use it all up.

The amount of solar energy available is staggering. For example, according to the George Washington Institute "All the energy stored in Earth's reserves of coal, oil, and natural gas is matched by the energy from just 20 days of sunshine".

*BPEC are one of the UK's foremost training organisations for renewable technologies.

http://www.bpec.org.uk

Why use renewable energy?

There are many UK Government schemes which offer financial incentives for adopting renewable technology.

Apart from the carrots, there's also a few legislative sticks in the pipeline as regards the required energy efficiency of new buildings which will "encourage" the adoption of renewables.

Governmental incentives aside, my experience is that for the most part, people and companies invest to reduce their energy bill - the environmental benefits a welcome by product.

Irrespective of the reasons to buy into renewables, those that do are reducing their carbon footprint which can only be a good thing.

Whatever the motive there is nothing wrong with saving money and as a return on investment it's almost certainly going to give a better return than your friendly high street bank - which is why some banks are investing in renewable technology.

If we consider other headline grabbing factors such as fuel security and fuel poverty, generating some free energy from natural resources is good for our environment as well as your pocket.

It's unquestionable that we are depleting fossil fuel supplies – it's only how long it will take that's debatable.

We have become energy hungry and continue to increase our energy demands almost daily.

If China's percentage of per capita car ownership equaled that of the USA it would consume more Oil than can be produced from every oil supplier in the World.

RENEWABLE ENERGY - A BUYERS GUIDE

Solar Thermal systems

If you've ever sat in a conservatory in the middle of winter on a bright sunny day, you've experienced the heating effect of light.

This is the principal of operation that most modern solar thermal systems work on. They capture energy from light rather than heat so they will work even when it's cold outside - as long as there is sufficient light.

Archimedes apparently managed to boil water as early as 214BC using a concave mirror so it's a pretty well established technology which has been somewhat refined since Archimedes experimented.

The main application of solar thermal systems is to heat domestic hot water which is then stored in a hot water cylinder.

However, if you have a seasonally used pool then solar thermal can be a good pool heating option and may well extend your season. In fact, a system that provides both pool heating AND domestic hot water is a very good combination.

How they work

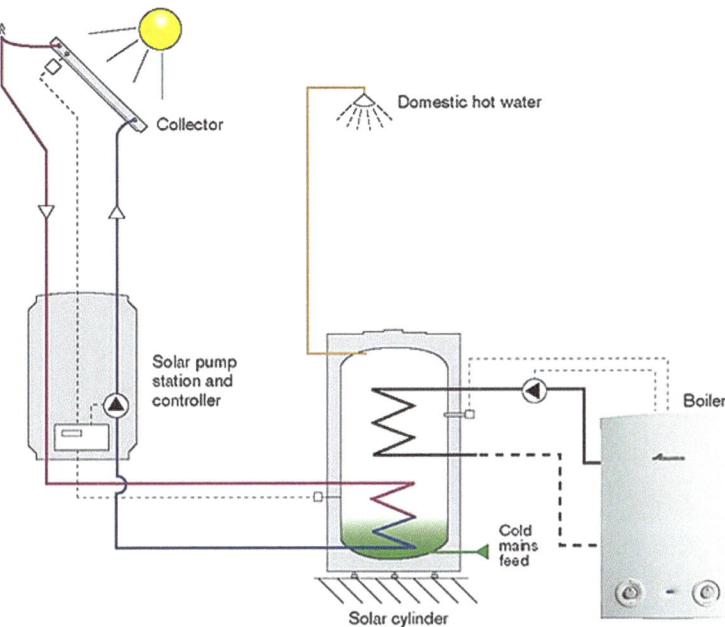

There are temperature sensors on the solar panel (AKA solar collector) and the water tank as shown by the two thin dotted lines. These temperature sensors are connected to a solar controller (the brains) and the controller compares the respective temperatures.

When the solar controller sees that the temperature of the solar collector is higher than the temperature of the cylinder, and hot enough to do some good, it turns the pump on and the heat from the solar collector is transferred to the cylinder. When this temperature difference is not large enough to contribute anymore heat, the controller turns the pump off and the cycle begins again.

As this example shows, in this case a boiler is included to provide heat when the solar systems contribution is insufficient. The boiler (or other supplementary heating source) also serves to boost the temperature on occasion to pasteurise the water and destroy any legionella bacteria.

There are two main types of solar thermal systems – Flat plate collectors or Evacuated Tube Systems.

Flat plate collectors are essentially two pieces of glass or perspex (plates) with pipe sandwiched in between.

A Flat Plate Solar Thermal Collector

Evacuated Systems consist of individual vacuum sealed tubes as in the picture below.

If you asked ten different suppliers which was best - flat plates or evacuated tubes - you would probably get ten different answers. However, overall system design and installation is far more important to system performance than whatever type of solar collector is used so don't base any decision purely on the type of solar collector.

Most customers tend to base their decision on the products aesthetics and in this respect Flat Plate collector seems to be the favoured option.

The solar collector/s (the panel) can be fitted onto a roof, into a roof, on the ground, wall - in fact pretty much anywhere.

As the amount of hot water and the temperature of the water produced will vary throughout the year, an immersion heater or boiler is usually incorporated to "top up" the hot water supply.

By virtue of the fact that a solar thermal systems performance (and therefore temperature of hot water) varies, it is necessary to have some other form of water heating rather than just solar.

In addition, to prevent legionella, (which causes Legionnaires Disease), it is necessary to raise the temperature of the cylinder periodically which kills off any bugs in the water.

On the other side of the coin, solar can produce very high hot water temperatures so it is also necessary to have a mixer valve fitted which blends hot water with cold as it leaves the cylinder so you don't burn yourself at the taps.

That said, regulations (in the UK) now require a means of stopping the solar thermal system "overheating" the cylinder to prevent excessively high hot water temperatures.

As you will need a cylinder (to store this solar hot water) you'll need sufficient internal space to accommodate the cylinder so bear this in mind. If you already have a water cylinder this will usually need changing for a solar cylinder as shown below. You'll notice this has a large coil at the bottom of the tank.

This is where the fluid in the solar collectors is pumped around which transfers heat to provide your domestic hot water. The coil is at the bottom so that all of the water in the tank is heated.

If this coil was at the top not all the water in the tank is heated as heat rises of course. This is called stratification which means "layers" - in other words you'd get hot water at the top of the tank and the water would get progressively colder the further down the tank you went. The use of two coils (as shown in the picture below) allow the connection of a boiler (or heat pump / wood burner etc) to the top coil so that both solar and boiler can heat the water as shown in the earlier diagram.

If you have to have a new cylinder as part of your new solar system installation, then go for a twin coil rather than single coil solar cylinder as the additional cost (for the second coil) is minimal and provides you with some future flexibility.

Also go for a twin coil cylinder that's been designed to accept a heat pump rather than boiler input. A cylinder made for a heat pump can be used with a boiler but a cylinder made for a boiler input can't generally be used with a heat pump so it's just a bit more flexibility and future proofing - more on this later. On the subject of cylinders, if you do need a new one it is well worth considering a thermal store (covered later) instead of the typical solar cylinder.

RENEWABLE ENERGY - A BUYERS GUIDE

A Twin Coil Solar Cylinder

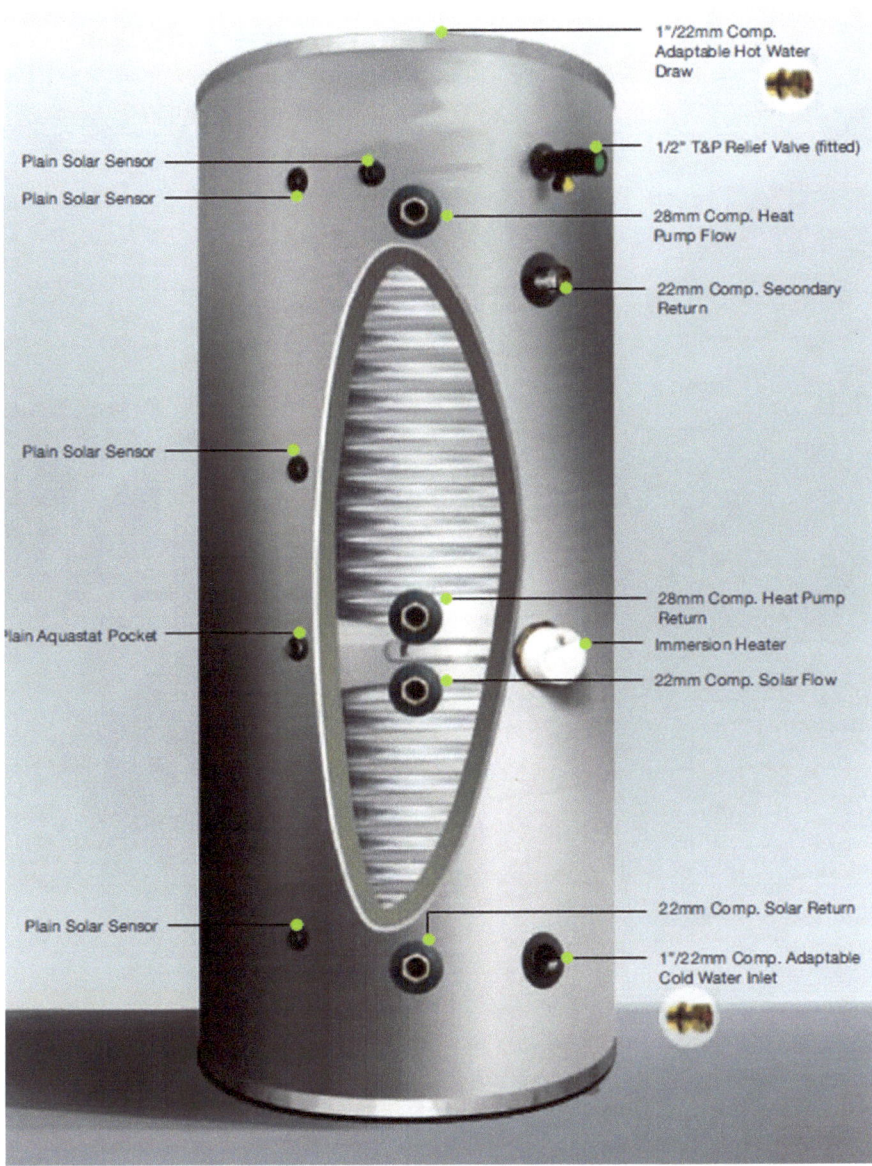

If you already have a cylinder with an immersion heater fitted you may not need a new cylinder at all.

There is a retrofit option which replaces your immersion heater - one such example is shown below. This may be slightly less efficient than a new cylinder but is a worthwhile consideration which should also reduce the system and install costs substantially.

Solar panels can also (in theory) provide some contribution to heating your home as well.

However, the amount of heat provided (in the UK) is generally very small and is not normally financially viable. Bearing in mind the greatest demand for heating is in winter which coincides with the time of lowest solar gain, the climate is working against us.

To make any worthwhile contribution to heating a substantial number of panels would be required and then you have the problem of what to do with all this excess energy in the summer.

This excess energy can be a real problem as your solar system can sit there boiling away damaging itself which may lead to other problems.

The exception to this solar contribution to space heating is a combined pool heating / domestic hot water system - assuming it's a seasonally used pool.

This is because we have a system that is sized for a pool (i.e much bigger than is required just for hot water), which in summer should also give us ample hot water, If we aren't using the pool in the winter we still have an "oversized" system that may well cater for all of our hot water needs, with some spare capacity to help the central heating. This set up could work well for a holiday facility.

When a system "boils" (known as stagnation) the fluid in it (called Glycol which is essentially anti-freeze)) is adversely affected and may need replacing so stagnation needs to be avoided.

One common method is to divert this excess heat into another circuit until it cools down sufficiently for the system to return to normal working conditions.

This "other" circuit may be as simple as a radiator which serves no other purpose than to dissipate excess heat. As the likely time for stagnation is in the height of summer this radiator will likely be somewhere out of the way.

The key thing to remember is that some means of stagnation prevention should be included in your system. *It is well worth asking your prospective supplier about this.*

In a typical solar hot water system expect around 50 to 60% of your domestic hot water to be provided by solar thermal.

Put another way, you may get all of your hot water needs from solar in the height of summer, whilst only a small fraction of your needs might be produced from solar in the winter. As we said right at the start, most modern solar thermal systems work on light rather than heat so yes they will work in the depths of winter, but of course the days are much shorter.

You might get 16 hours of good daylight in the height of summer but only 4 hours in mid-winter which makes solar thermal rather a seasonal producer and as mentioned, not really a viable contributor to space heating.

Facing which way?

South is good but South East or South West is almost as efficient. In fact they can even be orientated East / West and don't have to be perfectly inclined to your angle of latitude – flat on a roof or on a wall works pretty well and the solar collector/s can also be ground mounted.

However, whichever way they are orientated you will probably realise that shading is not good so bear this in mind with regard to the chosen location.

In any event your installer should check and take this into account and if in doubt should produce a sun shade diagram using a solar shade tool.

Maintenance

To cope with potential freezing Solar thermal systems are usually filled with Glycol (anti-freeze) and this will need checking every year or so and probably changing after around 5 years. This could costs anywhere from £100 upwards so you should take this into account.

On the subject of Glycol, if it stagnates it should be replaced in any event. After stagnation has occurred some types of Glycol can become toxic and may damage your system if it is not replaced.

Remember most systems also contain a pump which is subject to wear but generally these will last ten or more years.

Corrosion may be an issue with externally located components so warranty details should be checked as well as the longevity of the manufacturer.

There are two main other types of solar thermal systems - Thermosiphon and Drain back.

Thermosiphon you may have seen in countries with warm climates. They have a tank fitted at the top of the collector so everything's outside which is why they are not usually used in the UK as they're exposed to the elements and it's frankly too cold. On the basis that heat rises they don't need a pump to push fluid round which is the main advantage.

A typical Thermosiphon Solar System

Drain back systems are for some reason not very common in the UK and as such we won't discuss them here.

However, if a prospective supplier is promoting these don't discount them.

Planning and building control

Generally planning control is not required but check with your local council first.

Building regulations will apply and you should ask your installer about this as some installers are able to self-certify the work negating the need for building control notification.

The $64,000 question - Return on Investment / Payback

A typical household will use around 15 to 20% of it's total energy consumption on producing domestic hot water so this should give you some indication as to whether it's a good investment for you. That said, every household is different so what works for one won't necessarily stack up for another.

To accurately determine your return on investment you would need to know exactly how much it costs you to provide for your hot water needs but which of us does.

However, at the time of writing the UK has introduced the Renewable Heat Incentive - RHI. This may well make investing on solar thermal *(just one of the technologies that the RHI applies to)* a far better bet than it was pre RHI.

There is a non-domestic RHI scheme and a Domestic RHI scheme, both of which are administered by Ofgem: www.ofgem.gov.uk

The RHI scheme essentially pays you an amount for each unit of heat energy your system generates. The scheme applies to Solar Thermal, Air & Ground Source Heat Pumps, and Biomass.

The eligibility criteria are quite lengthy and the benefits of the scheme varies with the technology so it may be best to get your installer to explain the scheme (as it applies to you) and maybe ask them to do an economic viability analysis.

You can find out more at the Governments website at www.gov.uk

DIY?

It's quite possible for those with some good basic plumbing skills and you may make a substantial saving - you don't have to buy products with MCS approval which could save money but in the same vein neither will you get any incentives such as the RHI.

Those things said, if you do want to DIY then it may well be worth purchasing from a local installer (even though you will pay their marked up price) as you may be able to obtain a commissioning only service from them.

A well set up system is key for optimum performance.

Solar Thermal summary

Orientation is key so you'll need a southerly aspect available for the solar collector. The collector should also be unshaded, however some shading may be acceptable which will need to be assessed by your installer.

Around 3 to 4 square metres space is needed to accommodate the solar collector in a typical set up.

You will need sufficient space for a solar hot water cylinder and also an alternative method of water heating.

The economic viability will differ in almost every case and will also depend heavily upon which type of fuel you are displacing.

To obtain any Government grants or incentives the installer AND system must be MCS Accredited (Microgeneration Certification Scheme).

In the case of Solar Thermal systems only the Solar Keymark is an acceptable alternative to MCS product accreditation.

You can check whether the proposed products and installer are MCS accredited here: www.microgenerationcertification.org

Please remember that MCS Accreditation is not necessarily an indication of quality or workmanship.

Solar Photovoltaic

Enough energy from the Sun falls on the Earth in one hour to power the World's needs for a whole year.

Solar Photovoltaic (PV) systems produce electricity and are perhaps the most popular form of renewable technology in the domestic sector.

One of the main reasons PV really took off in the UK was / is the Feed in Tariff - a very generous *(perhaps too generous)* Government incentive which still exists but has been and will continue to be reduced. More on this later.

Roof mounted Solar PV and Solar Thermal collectors (centre)

If you don't like the look of Solar Panels on your roof a number of companies now make Solar PV Tiles in a variety of shapes, sizes and colours in an effort to "blend in" with the rest of your roof.

RENEWABLE ENERGY - A BUYERS GUIDE

Solar Photovoltaic Tiles

Solar PV systems capture the sun's energy using photovoltaic cells. These cells don't need direct sunlight to work – simply light – therefore they will still generate electricity even on cloudy days. Solar PV systems simply convert light into electricity.

PV is completely pollution free in operation with no complex controls.

It is also probably the most reliable source of sustainable power harnessing the most abundant energy source on our planet - sunlight.

PV does not need bright sunshine to work - the electrical output is dependent upon the intensity of the light to which it is exposed. If you shine a torch at a PV panel some electricity will be generated.

PV cells will tend to generate more electricity on brighter days but even on overcast days electricity will still be produced.

The performance of PV systems is very predictable, and with the data now available, a competent supplier should be able to tell you with a good degree of accuracy how much electricity your system will generate.

In the same vein they should also be able to tell you how much your system is going to save you on your electricity bills, and how much it could earn you from the Feed in Tariff.

The majority of PV panels work in the same way but there are three main types.

Monocrystalline panels are made from thin wafers of silicon. These are the most efficient type. The colour of these panels tends to be very uniform as shown in the picture below.

Polycrystalline works well but is less efficient than Monocrystalline.

As you can see in the picture below, the colour of Polycrystalline panels tends to have a "patchwork" look.

Amorphous silicon (sometimes called Thin Film) perform in lower light conditions, but you need more panel area to produce the same amount of power than other types as they less efficient - typically double the area of Monocrystalline for the same power output. You will probably have seen amorphous (or thin film) photovoltaic panels before - they are commonly used on solar garden lights, calculators etc.

As you'll probably have guessed, the more efficient the panel, the higher the cost although the price difference today isn't vast.

Other technologies are under development, perhaps of most interest is research being carried out using the photosynthesis effect that plants use.

Monocrystalline   Polycrystalline

All solar panels are rated in Watts Peak (Wp). As the name implies this is the peak amount of power (watts) that the panel will produce under optimum operating conditions.

Roughly speaking, a 1 metre square panel will produce about 100 watts per hour in good solar conditions.

What's a watt?

A watt is a unit of power - a 100 watt light bulb gives off more light than a 60 watt light bulb - it's more powerful

The 100 watt light bulb also consumes more electricity than a 60 watt light bulb.

If you look at your electricity bill you will see that you are charged in kilowatt hours. A kilowatt is simply 1,000 watts.

If we turn on our 100 watt light bulb for ten hours it will use 10 x 100 watts or 1 kilowatt of electricity. This is therefore 1 kilowatt hour (kWh) and which is what you would be charged by your electricity provider - 1 unit of electricity.

If we consider this in relation to a PV panel, if the panel is rated at our hypothetical 100 watts peak, this is what the panel will produce under ideal conditions and is the amount of electricity that you wouldn't have to purchase from the grid.

In reality, PV panels very rarely produce their maximum output simply because these "ideal conditions" don't happen very often.

Inverters

PV panels produce DC (direct current) electricity which is the same as you get from a battery.

Household appliances use AC (alternating current) electricity so we need an Inverter to convert DC into AC electricity if we are to use this power at home.

System types

Battery systems store the electricity produced so it's available day or night.

There is some loss of efficiency and battery life is finite.

You can use the DC electricity 'as is' for DC driven devices, or convert to AC with an Inverter for use with conventional appliances.

Battery systems are better suited to areas where a mains supply is not available.

One acquaintance uses just such a system to supply DC lighting and power a 'drinks fridge' in his remote garden. His drinks fridge is a 12 volt unit so he doesn't need an inverter.

There are a surprising number of DC products on the market now which have come about predominantly because of demand from the caravanning and boating market.

Direct Supply Solar PV Systems may be used to power products directly without batteries or Inverters. The solar panel is connected directly to a DC device - solar powered pond pumps are a good example of one such application. The speed of the pump will vary with the available light.

Grid Tie Systems (AKA Grid Interactive or Embedded Generation) are the most popular and most efficient. The main reason for the popularity of Grid Tie systems is because of the Feed in Tariff.

RENEWABLE ENERGY - A BUYERS GUIDE

Schematic of a Grid Tie PV system

1) Energy From The Sun Travels To Earth
2) Solar PV Panels Capture The Suns Energy
3) The Inverter Transforms The Energy (DC Current) Into Useful AC Electricity To Use In Your Home
4) The Generation Meter Records All Electricity Produced By The Solar PV System
5) The Free Electricity Produced By The Solar PV System Is Then Used By Your Household Appliances
6) Any Unused Electricity Is Then Exported

Grid tie means you (your installer) essentially connects up your PV system to your fuse board (consumer unit) which is of course also connected to the national grid.

In operation, essentially what happens is that your system generates electricity at a slightly higher level than that coming from the grid. What this means in practice is that electricity generated by your PV system will be used first and in preference to that supplied from the grid, with any shortfall being made up by the grid.

Examples

- Your system is generating 2 kilowatts of power and you are using 3 kilowatts at the time. 2 kilowatts will be taken from your PV system and 1 kilowatt from the national grid.

- Your system is generating 2 kilowatts and you are using 1 kilowatt at the time. This means that 1 kilowatt will be taken from your PV system and 1 kilowatt will be exported (sold) to the national grid.

This might seem wonderful - selling electricity back to the national grid and getting paid by the electricity suppliers.

However, in reality the amount they pay you is much smaller than they charge you for each unit of electricity.

At the time of writing UK electricity costs are around 15p per kilowatt hour, and for any electricity you export you will get paid around 4p per kilowatt hour - a difference of nearly 11p.

As you can see it makes sense **not** to export if you can help it, and offset more of the electricity you would otherwise use from the national grid. Fortunately this is quite possible and a number of companies now provide various types of equipment to achieve this.

One of the easiest ways to achieve this is to "dump" excess energy into an immersion heater thus reducing the work your boiler / heat source has to do to provide your hot water.

There are other systems, tips and tricks you can employ but these really need discussing with your chosen supplier.

On the basis that off-setting the electricity you take from the grid is better than selling electricity back to the grid, then some small lifestyle changes can help maximise the benefit you get from your PV system. Try and use appliances that have a high electricity demand during the day when your PV system is generating energy. If you're out during the day then you can use timers to turn on these energy users.

Increasing energy prices will actually make your system better value for money over time - the higher the electricity prices are, the more you are saving by off-setting your bills.

Which way should a system face?

The answer to this isn't as obvious as it might seem.

You night assume due South is best and for maximum generation it is. However, as most of a typical households demand tends to be later in the day, a westerly facing system will usually naturally offset more of your normal daily demand.

That said, if you adopt those minor lifestyle changes noted above due south comes out top.

However, a system orientated anywhere between south east and south west will still perform almost as well as due south.

The angle of inclination is also an important factor. The inclination of roof fitted panels will be dictated by the angle of your roof. Fortunately the pitch of most roofs is very suitable.

Just as with solar thermal systems it is most important to avoid shading the panels.

Connecting to the national grid

Your installer will liaise with your District Network Operator (DNO) to connect your solar PV system to the national grid. You may gather from this sentence that you can't DIY if you want to connect to the national grid - the system must be professionally installed by an MCS accredited installer using MCS accredited equipment (just as with the solar thermal system discussed earlier) to grid tie and benefit from the Feed in Tariff.

You can find out more about your local DNO here: www.energynetworks.org/info/faqs/electricity-distribution-map.html

System size

To connect to the grid with the least amount of red tape your solar PV system should be no more than 16A per phase, equivalent to 3.68kW.

This doesn't mean your chosen supplier will suggest a 3.68 kilowatt system - he/she may well suggest a slightly higher system size and 4 kilowatts seems to have become the common system size but be cautious - the previous common practise of installing more than a 16 amp per phase system is now carefully scrutinised - the below is an extract from www.yougen.co.uk - check out the Blog section for more information.

Extract

"For the purposes of changes that affect us, this means you can no longer have a Grid connected renewable power system that can export more than 16 Amps without special permission

Special permission in this case usually means more cost.

Please ensure your installer is familiar with this policy.

If your PV system is no more than 16A per phase (equivalent to 3.68kW) it falls under fall under G83/1-1 Stage 1, and your installer can simply inform the DNO within 28 days of commissioning that a connection has been made.

You can have a larger system than this installed but there are additional requirements and you / your installer will need to involve the DNO pre installation. Additional connection costs are likely.

Grid tied systems - What happens when there is a power cut?

If you have a grid tied PV system it will automatically shut itself down. This is part of the requirement for an Inverter to achieve G83/1-1 certification.

The reason for this is that in a power cut, your system could continue generating electricity sending it back out onto the power cables. If an engineer is working on the cable lines, the electricity you're generating might just upset the engineer if he decides to grab onto the cable - in other words it's a safety issue so your own system has to shut down.

There are ways your system can continue to operate in a power cut but that really is something to discuss with your chosen supplier.

Planning permission

Pretty much the same as for Solar Thermal systems. Usually comes under permitted development but your building control office might want to check and advise as to whether your roof structure is suitable - your installer should be able to advise and liaise for you.

The safe thing to do is ALWAYS check with planning and building control first.

The Feed in Tariff - FIT

If you don't have an export meter fitted (an electricity meter which measures the amount of electricity you send back to the grid), you are deemed to be exporting 50% of the electricity your system generates back to the grid. Their FIT payment will be based on this amount no matter what their actual export is. It is quite normal not to have an export meter if your solar PV system is less than 30kWp. You can however choose (and pay for) an export meter to be fitted but do your sums first. It usually only pays to have an export meter if you have a system greater than 30 kilowatts.

As part of their job, your installer will register your site with the Microgeneration Certification Scheme, and you will get a certificate which you can use to claim Feed-in-Tariff.

As these Feed in Tariffs are, and will continue to change, then you will need to obtain the appropriate advice from your installer. However, even though the tariffs have been reduced at the time of writing, it is still a good deal in my opinion and your installer should be able to provide you with calculations as to the economic viability of investing in a PV system.

You can get an idea of how much revenue and savings your system might benefit you by using the Solar Energy Calculator which is available at www.energysavingtrust.org.uk

As a very rough rule of thumb a typical 2 kilowatt system will meet about 30% of the average households' electricity needs.

Free Solar PV

Something for nothing - or nearly.

Essentially a company will install PV for you free of charge on the basis that they receive the Feed in Tariff while you get to use the electricity their system is generating.

The length of time this applies for may vary from supplier to supplier and the electricity you use from the onsite generation will not always be completely free but may be at a discounted price.

Recently, for the supplying company to benefit from the maximum Feed in Tariff your property needs an EPC (Energy Performance Certificate) in Band D or higher.

If you are considering a free PV offer you should check with your mortgage provider (if applicable) that they have no objections to such an installation and you should work out the annual benefit to you.

You should also bear in mind that if you sell the property, as the PV agreement is tied to the house, your purchasers may be forced to inherit the deal. Not necessarily a problem but something you will need to consider.

Home insurance

Do check that your insurance cover includes cover for a PV system - PV panels are a valuable commodity and one worth protecting. If you have "free PV" the supplying company will normally own the system outright so really they should be insuring the installation but as always, check the agreement.

System lifespan

PV systems are completely passive in operation with no moving parts so really there's nothing to wear out. It is expected that today's PV panels will have a lifespan of 30 or more years.

However, the amount of energy generated by a panel will tend to slightly reduce over time.

The specification will normally state that the panels will still produce 90% after ten years or something similar, and in reality they'll probably still be going strong in year thirty.

Wind Turbines

Wind Turbines seem to fall under the bracket of Marmite products - you either love them or hate them, and they seem to generate as much controversy as they do electrical energy.

They can be used to do more than generate electricity and are used for pumping and of course milling flour - windmills - which IS NOT the same as a Wind Turbine. You'll probably gather from the windmill reference that the concept has been around for a long time - it's just been refined and developed over the years.

As turbines attract so much publicity and the number of wind farm developments increases I don't think it's necessary to explain what Wind Turbines are.

However, it's worth mentioning there are mainly two types, Horizontal axis (HAWT) and Vertical axis (VAWT). There are pro's and con's to both but in the main and for the purpose of this book, we'll stick with horizontal axis turbines which are far more common and more readily available.

Vertical Axis Wind Turbine

RENEWABLE ENERGY - A BUYERS GUIDE

Horizontal Axis Wind Turbine

At the time of writing the UK had 359 wind farm sites with 4035 machines producing 7.4GW (gigawatts) which is approximately 5.2% of the UK electricity supply.

UK Wind Resource

This link is pretty much real time wind speed date as it happens: www.xcweather.co.uk

The website www.rensmart.com/Weather/BERR gives some good NOABL data (Numerical Objective Analysis of Boundary Layer). The data is slightly different from that offered by the met office. NOABL data breaks down the UK into 1km grid squares and gives an estimate of the wind speed at the average height of the square. It makes some assumptions and tends to overestimate wind speed in valleys and under estimate wind speed at hilltops.

You can also find a useful wind speed prediction tool at www.energysavingtrust.org.uk

Remember, wind speed data tends to be given at different heights on different sites. As a rule, the higher the better as turbulence is reduced with height, and the difference in wind speed at say 5 metres and 10 metres can be quite dramatic.

The height of a turbine will directly impact on its performance.

The most important wind data you need is that which is specific to your site. Whilst you can get some pretty good data from the above (and other) websites, the only real way to be sure is to buy a wind logger - really a weather station that records data for you.

For this data to be of use the wind logger monitor needs to be fitted in the intended position, and at the proposed height, of your wind turbine. You really need a years' worth of data for this to be a meaningful exercise.

A typical wind speed data logger

Location, location, location

This is key to how well (or not) a turbine will work.

Wind turbines prefer a nice steady airflow called laminar airflow.

Any turbulence from buildings, trees, hills or any other obstructions, will disturb the airflow and therefore affect the performance of your turbine.

This is particularly important for grid tied wind turbines, as the inverters have to synchronise themselves with the incoming mains supply for period of time a before they connect to the grid.

If there is significant turbulence this will affect the ability of the turbine to connect and remain connected to the grid.

It's worth mentioning that Solar PV has to go through this same synchronisation but the amount of light doesn't tend to vary quite so dramatically as wind speed so it's less of an issue for PV.

How windy is windy?

The Beaufort wind force scale can be found at the met office link here: www.metoffice.gov.uk

This gives a nice clear explanation of how windy is windy.

How much wind do you need for your turbine to work well?

Here are some very approximate rules of thumb based on a 10 metre turbine height. Wind speeds are given in m/s (metres per second).

0 - 3.5 m/s ; of no use

3.5 - 5 m/s ; may be usable for a very small standalone system

5 - 6 m/s ; considered sufficient to make a FIT turbine worthwhile

6+ m/s ; good wind farm potential

Applications

These are the same as for Solar PV although generally speaking very small wind turbines (less than say 500 watts) tend to be used for charging batteries with perhaps the marine market being the main consumers of such turbines.

There are also "wind turbines" that can be used in the water either driven by the waters flow or towed behind a boat so the marine market is a "buoyant" one. They tend to be called water or hydro turbines but they are essentially wind turbines which just happen to be water powered.

Again as with Solar PV, turbines can be grid tied in order for you to obtain the Feed in Tariff although the tariff rates differ from those for PV. Bearing in mind the need for a wind turbine inverter to synchronise to the grid as noted above, many people tend to prefer battery systems which are not affected by this synchronisation period and ON / OFF cycle.

As with Solar PV grid tied systems, if the wind turbine is up to 16A per phase (equivalent to 3.68kW) it falls under fall under G83/1-1 Stage 1, and your installer can simply inform the DNO within 28 days of commissioning that a connection has been made.

Anything bigger means contacting your DNO just as with PV.

Power ratings of wind turbines

This is something you really need to be aware of when looking at what the market has to offer.
Turbine manufacturers give a rated power (kilowatts) to indicate their output, however, this tends to be given at different wind speeds measured in meters per second (m/s) making it very difficult to make a direct side by side comparison.

If you have two wind turbines rated at 1 kilowatt but one is rated at 8 m/s and the other is rated at 11 m/s then it's not comparing apples with apples.

To overcome this problem Renewable UK introduced the Small Wind Turbine Performance and Safety Standard.

As part of the MCS approval process the wind turbine has to be measured against this standard which should help you assess the suggested output when comparing wind turbines.

This doesn't mean that turbine A is better than turbine B - you just have to be aware of the power curves.

Having conscientiously studied and measured your wind speed at site, then reference to a turbines power curve should help you choose a turbine designed for a particular average wind speed.

Turbines tend to be designed to perform most efficiently at a particular wind speed chosen by the manufacturer so it is "horses for courses".

The Small Wind Turbine Performance and Safety Standard information generally includes:

- The BWEA Reference Power, which is the rated power of the wind at 11m/s

- The BWEA Reference Annual Energy, which is the amount of energy in kWh that the turbine will produce in a year at a constant wind speed of 5m/s at a stipulated hub height

- The BWEA Reference Sound Levels at 25 and 60m rounded up to the nearest decibel (dB) from the turbine.

A turbine which meets the requirements of this standard will have a label stating "Certified by BRE".

Maintenance

A Wind Turbine does of course have moving parts so some maintenance will be required.

Security of the fixings should be checked frequently. On some turbines the cable runs up the centre of the turbines mounting column and may become twisted so will need "unwinding - it depends on the connection method whether this situation occurs.

A visual examination of the turbines blades and other mechanisms is also recommended.

It is sometimes possible to see if the turbines blades have become unbalanced or twisted when the turbine is spinning.

None of this is too demanding but these are just things to keep an eye on. Your supplier should be able to provide advice on your particular turbine.

Myth or fact?

Wind Turbines are perhaps one of the most controversial forms of renewable energy there is and as such tend to be the cause of much debate. They don't work when it's too windy, they kill birds and bats, they're noisy etc.

Most of this is just myth but you need to be comfortable with your choice so after some research you just have to take a view and draw your own conclusions.

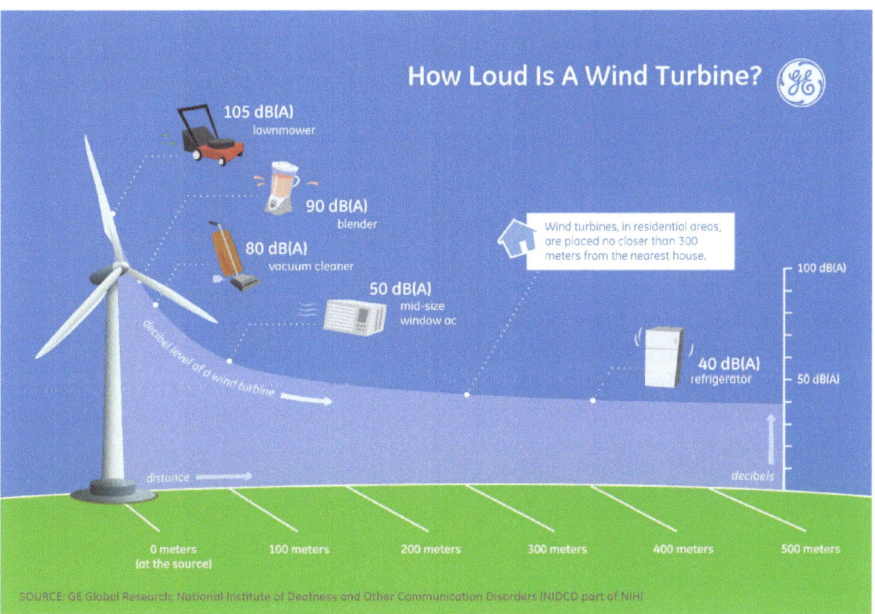

Planning

Whilst the laws regarding planning permission have become more relaxed you should always check with your planning office and it never hurts to involve your neighbours to avoid the NIMBY situation getting out of hand. Building regulation approval is usually also required but if grid connecting this is a job for your installer.

Micro Hydro

In terms of electricity generation and efficiency, micro hydro is about as good as it gets.

The downside is that Hydropower is very site specific and being dependent upon rainfall may mean that you have much less power in the summer than the winter.

In fact, the amount of available / usable water flow during drier periods tends to be a key factor in determining the viability of a micro hydro installation.

The two key factors in assessing the viability of a hydro scheme are water head and water flow.

Fortunately rivers flow downhill which means they originate at a higher point so it has potential energy because of its height - in hydro terms this is the "head" of water.

The greater the height (head) and amount (flow) of water the greater the amount of electricity that can be generated.

These are the two key things that need to be assessed for a proposed hydro system.

As the flow will vary throughout the year it is important to select a turbine that will generate power throughout the year. If you choose too large a turbine it may be producing a good amount of power in the "rainy season" but may sit idle for long periods in summer.

As a rule of thumb turbines are sized to operate on a maximum flow equivalent to the average daily flow.

There are a myriad of different types of hydro turbines, each suited to a particular head and flow rate so the available water source will tend to dictate which type of turbine will do the job.

Just as with wind and solar power it is permissible to grid tie a hydro system and obtain the feed in tariff.

Hydro is a bit of a black art and is very much a job for a professional with a myriad of different types available - each having their own benefits and application.

Suppliers of hydro turbines should be able to outline what head and flow their particular turbine/s needs and should be able to point you in the right direction.

Hydro turbines aside, we shouldn't forget Water Wheels which have been around for a long time - in fact the Romans used water wheels in mining projects.

Water wheels were still in commercial use well into the 20th century, but they are no longer in common use.

Whilst water wheels were used extensively for milling flour and grinding wood into pulp for paper making, the main application for those in existence today is producing electricity.

You will almost certainly need planning permission. As all water courses in the UK are controlled by the Environment Agency you will almost certainly need their permission in the form of an abstraction license so that is probably the place to start.

As Micro Hydro is a specialist subject I won't expand upon it here but it really is worth exploring if you think you may have a viable water source.

Micro Hydro in action

These pictures shows the "business end" of a Micro Hydro installation supplied by Brew & Corkill, Isle of Man.

The generator itself (the round white thing with the pulley) is from a wind turbine, modified to suit a hydro set up.

For the purpose of the photo the protective covers have been removed.

The generator is only circa 2 kilowatts but this is a reliable water course so the expectation is that some 36 kilowatts of electricity will be produced every day, 365 days a year.

To put this in context the owners are unlikely to ever pay another penny for light, heat or power - an attractive proposition.

This system is off grid - the property has never had mains power and has always used a diesel generator - so a large battery bank stores the power for use as and when required.

A great use of a natural resource.

Micro CHP (Combined Heat and Power)

Established in the commercial world but a relatively new technology to the domestic market which has only enjoyed a fairly small take up to date. The idea is that you can heat your home and generate electricity at the same time.

Essentially it is a boiler (usually mains gas or LPG) that generates heat and electricity at the same time, from the same energy source. In fact, most of the models currently on the market look just like a standard boiler.

As the main domestic requirement is for heat, the amount of electricity produced is only about 1/5th of the heat output. In other words you'll get a heat to electricity ratio of around 5 to 1.

A typical domestic system will give about 1kW of electricity. Of course the amount of electricity generated depends on how long the system runs.

That said, just as with Solar PV and Wind power, systems can be grid tied so any electricity you don't use can be sold back to the grid.

It's debatable whether you could call this a renewable technology being powered by mains gas or LPG, but, it could be reasonably considered to be a low carbon technology which may be more efficient than just burning gas and using electricity from the national grid.

That said, no doubt we will see these systems using fuel sources derived from natural resources in the future (rape seed oil, biomass, etc.) when we will be able to class it as a true "renewable technology".

It's still early days for this technology so the jury's out for now.

Mixing Technologies

As your energy demand tends to be higher in winter than summer, and bad winds tend to accompany bad weather, a wind turbines generation does match the annual demand more closely than perhaps solar.

However, on the premise that it's sunnier in summer and windier in winter then a mix of solar and wind products provides a good balance. You'll probably have seen street lights that have a very small wind turbine and solar PV panel for just this reason. This combination tends to be very good for battery charging and off grid systems.

Small wind turbine and solar PV panel - both connected to a battery bank

Also notice the wind data logger just to the right of the turbine

PVT (Solar Photovoltaic / Solar Thermal)

More recently PVT has become available. This essentially combines a solar thermal panel with a solar photovoltaic panel. It could be biased towards either thermal or electrical (PV) energy production but in the main is principally PV.

Overall the efficiency of such systems has tended to be less than buying the technologies separately but the concept is excellent and advantageous for PV output in particular.

This is because PV produces more output at lower ambient temperatures as the picture below demonstrates.

The picture below shows a meter connected to the Solar PV panel in the above picture.

As you can see it is reading 111 volts and yes - that's snow in the background. During summer the average voltage reading was around 90 volts.

Of course the amount of solar irradiation (solar power) is highest in summer which is when the ambient temperatures are also highest which works against the PV energy production. So, by capturing this PV heat energy using solar thermal we can essentially cool the PV panel down enhancing its efficiency and as a double whammy get some free heat energy at the same time.

An added bonus is that you can collect more energy from a smaller amount of space than would be required if you put in separate solar thermal and solar photovoltaic systems.

As regards the UK market this combined technology may qualify for both the Feed-in-tariff and Renewable Heat Incentive so it is a contender.

The market for such systems is immature at present and there are only a limited number of suppliers but this will grow as new players enter the market.

I've no doubt the efficiency of this technology will improve significantly and it is likely to become mainstream in the near future.

Space Heating

Some 50 - 60% of our energy costs are incurred by space heating so it makes sense to try and reduce this amount before (or at the same time) as improving your heating system.

The main ways of doing this are:

- Improving insulation - the roof space is the easy one but not the only area - floors can also suck up a lot of your heat, as can walls.
- Reducing loss through windows and doors - double glazing is the one that will spring to most peoples mind.
- Reducing ventilation losses. These are mostly those badly fitting windows, doors, air leaks through letterboxes, chimneys, fireplaces, loft hatches and so on. You can arrange to have your home tested for air tightness or wait for a windy day and wander around with an incense stick (which will also make your home smell nice). Most people would be surprised at how badly their homes are sealed.

The above measures may not be exciting but generally speaking these are "one offs".

A lot of the improvement measures could be done by a competent DIYer at relatively little cost so before improving your heating system spend a little time ensuring what heat you do produce stays in the house.

Space Heating Emitters: Standard Radiators

Even something as simple as standard radiators can be improved at relatively little cost. Those good old single panel radiators aren't the best and just changing a single panel for a double panel radiator with double convector's can make a huge difference.

If you refer to the sketch below you can probably see that changing a single panel radiator for a double panel double convector radiator won't necessarily mean changing the pipework.

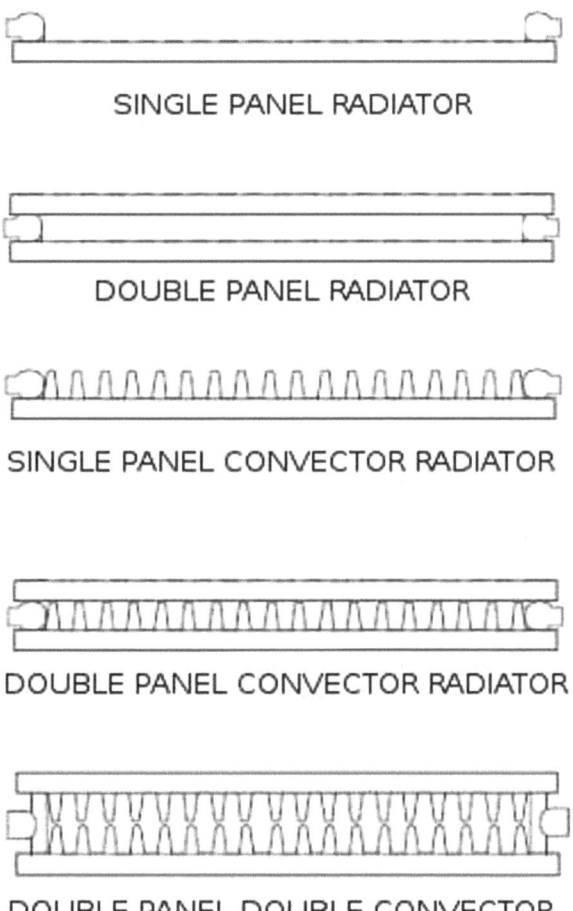

SINGLE PANEL RADIATOR

DOUBLE PANEL RADIATOR

SINGLE PANEL CONVECTOR RADIATOR

DOUBLE PANEL CONVECTOR RADIATOR

DOUBLE PANEL DOUBLE CONVECTOR

The convectors shown do just as they say - they help convect heat around a room.

Changing a single panel radiator for a double panel double convector radiator will more than double the surface area of the emitter*, (which is a radiator in this instance), so you can reduce the emitter temperature and still get the same amount of heat in the room.

If you can reduce the temperature needed to "run" your radiators then your heater need not work as hard thereby reducing your running costs - whatever the heat source.

Heating engineers tend to use the generic term emitters to describe all heat output devices.

Space Heating Emitters: Smart Rads

Smart rads seem to be the title that these products have been tagged with which is why I've used it here although LST (Low Surface Temperature) tends to also be commonly used. They are not necessarily the same thing but the principals are, which is increasing the output of a radiator using a lower flow temperature whilst keeping a small footprint.

Smart Rads are the "in thing" where radiators are concerned and with good reason.

These "Smart rads" operate on lower flow* temperatures (hence the descriptive LST tag) than a standard radiator and can be as low as 35ºC - a standard radiator might typically operate on a flow temperature of 65ºC or more so there is a big difference between the two.

Smart rads generally contain a lot less water than a standard rad which means they're quicker to react than standard rads. This combination of less water and lower running temperatures means they consume less energy whatever heat source you are using and they are ideal for use with heat pump systems.

Some contain a small fan which helps circulate the warm air around a room which increases effective heat output.

Some products also contain their own thermostats and programmable timers so each room could be programmed individually, potentially reducing energy consumption even further.

Lastly, because the working temperature is lower than a standard rad they can be used safely in hospitals, schools or anywhere that high temperatures might lead to risk of burning.

You might wonder why these "Smart rads" have not made standard steel radiators obsolete. The answer is simply price - there is a lot more to a Smart rad than a lump of pressed steel but it's generally only a one off investment so it warrants consideration.

Flow temperature is the temperature of water leaving your boiler. Return temperature is the temperature of the water going back to the boiler / heat pump / heat source.

Low Surface Temperature Radiator (LST)

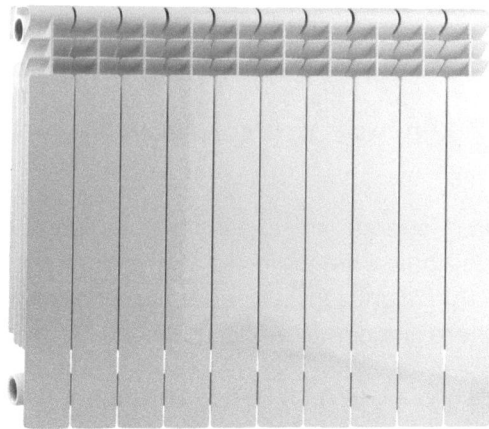

Space Heating Emitters: Underfloor Heating (UFH)

Thought by many to be the most comfortable and efficient (for wet UFH) form of heating emitters and warm floors are a very nice thing to have.

Underfloor heating is not a new concept as it was first used by the Romans who understood the benefits of invisible 'radiant' heating.

There are electric UFH systems and wet UFH systems. Wet systems are simply pipes embedded in the floor through which water is pumped around as with radiators.

Underfloor heating can be fitted under virtually any type of floor finish. However, different floor coverings have different levels of thermal resistance or insulation.

These values are known as the 'R' or TOG value - also used for continental quilts.

The most suitable floor coverings are those with low TOG values such as tiles or stone for example. Generally speaking the harder or denser the floor covering the lower the TOG.

The one downside of under floor heating has been that retro fitting is not really a viable option (unless you intend ripping up the floors anyway) so has generally only been suitable for new build or renovations. However, some manufacturers now produce "wet" systems that are very slim in profile and are designed to be laid on top of your existing floors offering a retrofit possibility.

Traditional radiators quickly heat the area in the immediate vicinity around them, with the heat rising and slowly moving around the room. An underfloor system heats from the floor upwards at a lower temperature and the emitter are is large as it is pretty much the whole floor.

Essentially your floor becomes the emitter which is one of the perceived downsides of UFH. Because a large emitter area (the floor) has to be heated it tends to be slower to warm up (and cool down) which means it takes longer to feel the warmth. This is true but as we are aware of this it is simply a matter of setting the controls (timer) to come on a little earlier than you might if you were using radiators for example. However, modern optimum switching heating controls (covered later) can take care of this delayed heating response automatically.

Under floor heating pipework just prior to screeding

The main reason wet UFH systems are considered more efficient than the alternatives is because they can be run at much lower temperatures than traditional radiators, meaning your heating system doesn't have to work so hard and will use less energy.

An under floor heating system might typically be run at just 35°C whilst standard radiators are generally run at around 70°C.

The larger the heating emitter, the more economical the room heating will be.

Because heat is radiated from the floor up, the overall room temperature can be 1 to 2°C lower than with conventional radiators, which by itself can give energy savings of 5 to 10% which is well worth having.

With wet under floor systems chilled water can also be pumped through the pipework thus providing heating and cooling.

Heating Controls

The importance of heating controls is sometimes overlooked but they are vitally important if you are to get the most out of your heating system, AND, run it as economically as possible.

Some heating systems have almost no control at all - relying instead on a thermostat in a boiler which senses the temperature of the returning water to turn itself off.

Some boilers are also fitted with a time clock which gives a slightly improved degree of control.

Such limited control is about as simplistic as you get get which means a system is far less efficient than it could be. The simple addition of a room thermostat to control the boiler would have a significant effect upon general comfort levels and system efficiency.

Heating Controls: Thermostatic Radiator Valves (TRV's)

These are as their name suggests - a thermostat which is incorporated into the radiator valve.

They're usually fitted in the most aesthetically pleasing position which is vertically orientated at the bottom of the radiator. This is also about the worst place and position you could choose.

However, they still serve a purpose and you have control over each radiator / room in which they are fitted.

A typical Thermostatic Radiator Valve – TRV

Heating Controls: Programmable TRV's

The same principal as a standard TRV but programmable which allows users to control temperature and time for each room / radiator that has one of these fitted.

Each radiator can be controlled in an extremely flexible way. For example, a bathroom could be programmed to a cosy higher temperature in the morning, a low temperature in the day to keep things ticking over, and a comfortable economy temperature in the evening. Similarly a lounge could be heated just weekday evenings but all day and evening at weekends.

This is about as flexible as you are likely to get and is like "zoning" each room in the house whilst achieving a high degree of heating control efficiency.

Hysteresis, Optimised heating controls and the Internet

Hysteresis in a heating sense refers to where changes in one part of the heating system take time to have an effect on other parts.

If we consider this in relation to room thermostats for example, a clunky old thermostat (the type that goes click when you turn them) might be set to let's say 20°C. However, the thermostat might only be accurate to plus or minus 2°C so might not turn off your boiler until the temperature is at 22 degrees and not start up again until the temperature has dropped to 18 degrees. In other words we have an operating band of 18 to 22 degrees which is a very large temperature swing, uncomfortable and inefficient.

Because of this large hysteresis (and our internet controlled lifestyles) Optimised and intelligent switching is becoming more popular.

Essentially these are very accurate thermostats (maybe just plus or minus 1/2 a degree) that also learn your lifestyle and homes heating response. Check out www.nest.com for one such example.

I guess no modern heating control system would be complete without Internet enabled control which can do more than just impress your neighbours. As we know there's an *"app for everything"* and heating control is no different.

Internet control can give you the facility to control and program your heating remotely via your smart phone.

A word about Condensing Boilers

This isn't a book about boilers but it is worth noting something that many (including heating engineers) aren't aware of.

It's a legal requirement (in the UK) that newly fitted gas and oil boilers must be condensing boilers. These recoup heat that would usually go to the outside world via the flue.

Condensing boilers are therefore more efficient than non-condensing boilers, but, and it's a big but, a condensing boiler cannot condense unless the return temperature is less than 54ºC.

Many boilers tend to be set to something like 75ºC (flow temperature) and as a typical heating system has a return temperature of around 10 degrees less than the flow temperature you can see we don't get anywhere near this 54ºC figure.

In fact, boilers don't tend to condense at their most effective until the return temperature is around 35ºC.

So, if you can turn your boilers temperature down and maintain a comfortable level of heat, then it's worth doing. Generally speaking lower radiator temperatures are more comfortable anyway.

Thermal stores

A lot of heating systems use some sort of water cylinder for either the hot water, central heating, or both.

Combi boilers were introduced to eliminate the space needed to accommodate a water cylinder but as with all things, there are pro's and con's to every approach, although now that water cylinders have better insulation than Dad's old coat, airing cupboards are sadly a thing of the past.

However, if you do have space for a cylinder the current trend seems to be Thermal stores.

Essentially these gives you the ultimate in future proofing and flexibility as most thermal stores will accept inputs from heat pumps, solar thermal, boilers, wood burners, direct electric and pretty much any other form of heat source you can think of.

These heat sources don't have to be fitted from the start but can be added later so that's the flexibility angle covered but what is a thermal store?

Some might think it means a very large cylinder (maybe 2,000 litres or more) to store heat energy - i.e. thermal store - and this would seem a fair definition. There might be a number of reasons for wanting to heat and store a large volume of hot water - the use of "cheap rate / overnight" electricity for example. With log boilers it is common to fill them up and light them once a day to heat large "thermal stores" so you have sufficient hot water to get you through the day until you next load up the log boiler.

However, whilst this is a fair definition, it is not the one that is most commonly applied to thermal stores which can be as small as 300 litres. The main characteristic of a thermal store is the ability to use multiple heat sources and supply instantaneous domestic hot water.

Who knows, in a few years we might be popping in to our local DIY shop to pick up a hydrogen fuel cell to plug into our thermal store so a bit of future proofing never goes amiss.

RENEWABLE ENERGY - A BUYERS GUIDE

An example of a Thermal Store

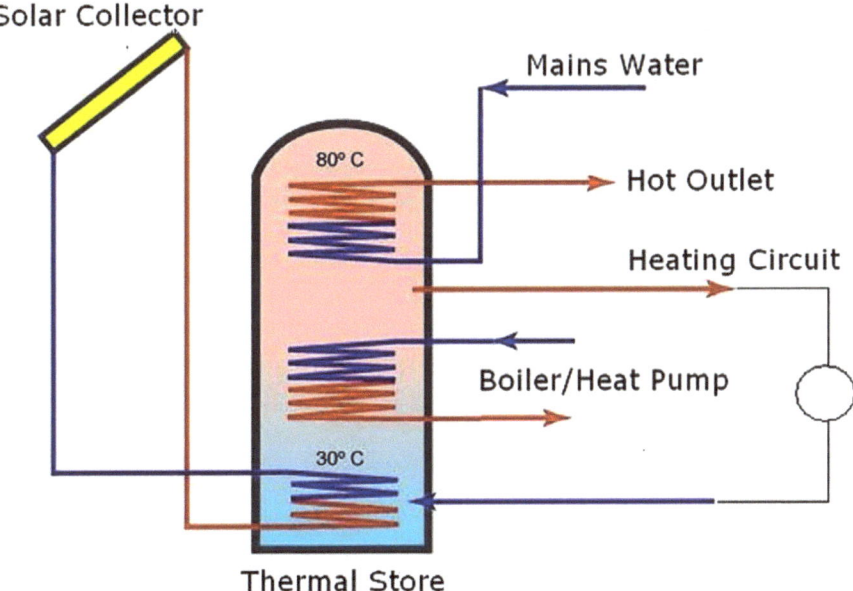

This example shows a solar thermal input and either a boiler or heat pump input.
However, a thermal store might have the facility to accept many more inputs allowing the possibility to introduce a new heating source at a later date.

In this example the domestic hot water is passed through a large heating coil.
As this coil is "sitting" in very hot water - 80°C in this case, the cold water that enters the bottom of the coil is hot by the time it leaves the cylinder. This is a common characteristic of thermal stores.

Heat Pumps

The current vogue form of property heating and with good reason.

Heat pumps are considered renewable technology because the heat they extract from the air, ground, or water is constantly being renewed naturally which give them their green credentials.

The simplest way to think of a heat pump is as a super-efficient type of boiler that can provide heat for radiators, under floor heating and domestic hot water.

Being relatively new to the market they are not, as yet, widely understood. However, the laws of thermodynamics (the principals of heat pump operation) is probably not the reason you're reading this so let's demystify this technology by looking at some basic principles.

One of the most common analogies used for heat pumps is "You've probably got one in your home" - AKA the fridge.

Well, that's true - the principals of a fridge and the increasingly popular heat pump are the same.

To explain; to produce a lump of ice we don't cool water down as such - we extract the heat from it to a point where it becomes a solid. So, put your ice cube tray in the freezer and the "heat" extracted from the water is ejected into your kitchen until we get ice.

That's why the back of your fridge / freezer is warm and also the reason why you should have some space all around the fridge so you don't impede this dumping of heat. The "heat" from the water is literally pumped out of the fridge - a heat pump.

Conversely, when you take the ice out of the freezer it melts because it is absorbing heat which turns it into a liquid. In fact, the ice melting would cool down your room but by such a minuscule amount as to be imperceptible.

So essentially we are taking heat from one place and putting it into another which is what a heat pump does - it just happens to "magnify" the heat it's extracting from the air, ground, water etc.

Let's define the term heat.

When we think of heat we tend to think of it being 20 plus degrees but heat is a relative term. If you stood outside and the temperature was say zero, then came into a room where the temperature was say 10 degrees it would feel "warm" - at least for a while.

Heat pumps have a refrigerant in their system - typically they use refrigerants called R134a which has a boiling point of −26.3 °C or R410a which has a boiling point of −48.5 °C.

As we said, the term "heat" is relative so our hypothetical outdoor temperature of zero is positively tropical to the heat pumps refrigerant circuit, which is why you'll see (air source) heat pump specifications that state things like *"will work even when the outdoor temperature is as low as - 15 degrees"*.

Back to our ice cube: If you put our ice cube into a kettle it would first turn to water and then to vapour (steam), in other words it changed state. This is what we do to the refrigerant in a heat pump - we make it change state from a liquid into a gas and vice versa.

When our refrigerant changes state from a gas to a liquid it gives of heat just like our ice cube changing state from water to ice.

When our refrigerant changes state from a liquid to a gas it absorbs heat just like our ice cube changing state from ice to water.

A gas can hold a substantial amount of energy.

Next we need a method to make our refrigerant change state and increase (magnify) the heat to a level useful for heating our home and hot water. To make our refrigerant change state from a liquid to a gas we simply have to make it hotter than its boiling point (which as we've seen is pretty low). The method we use for this will vary so we'll discuss the "how" when we look at some different types of heat pump.

However, the method used for magnifying the heat is compression which is what our heat pump mainly uses electricity for - to power a compressor.

You've probably all used a compressor at some time - an example of a compressor is a bicycle hand pump. Put your finger over the end of the pump and you are compressing the air inside the bicycle pump and in doing so are creating energy. Do this a few times you will feel the body of the bicycle pump getting quite warm so it's a pump that creates heat - a heat pump.

The pumps inside todays "heat pumps" are somewhat more sophisticated than a bicycle pump but the principle is the same.

The working cycle of a heat pump

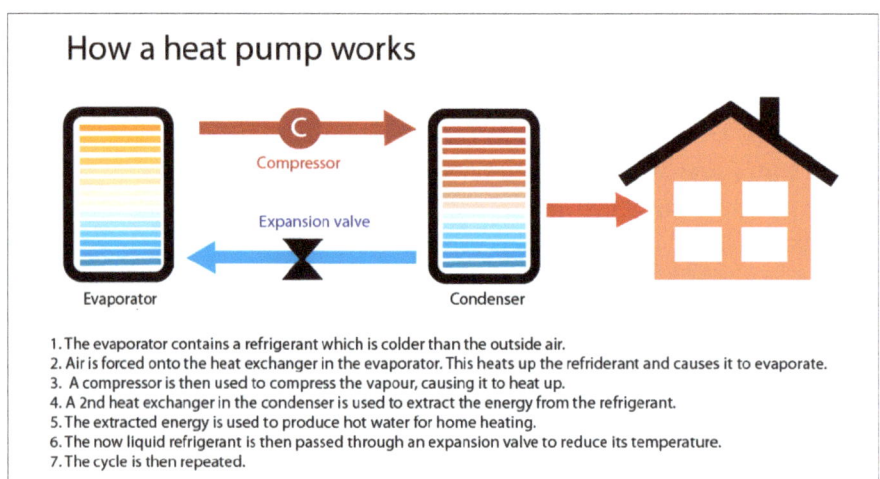

Diagram kindly supplied by Lailey & Coates

Whilst heat pumps can use different methods to produce (magnify) heat, the most common method is known as the "Vapour Compression Cycle".

Let's consider an Air Source heat pump, or more specifically an air to water heat pump.

Looking at our diagram above - we pass the refrigerant through an evaporator (very similar to a car radiator) which is outside.

Let's say the outside temperature is zero - bearing in mind the boiling point of refrigerant, zero is absolutely tropical to the refrigerant so it turns from a liquid into a gas with more energy than it had before. It has extracted energy from the outside air.

With the aid of a compressor we magnify this "heat energy".

To extract this heat we now pass the refrigerant through a condenser where it turns from a gas back into a liquid (like steam would on a window) and gives off its heat.

To start the process over again we just need to "uncompress" the liquid using an expansion valve and the cycle starts over again.

Now we've covered Thermodynamics 101 let's look at some heat pump specifics.

Why use a Heat Pump?

Because they can have efficiencies of greater than 100%?

The efficiency of heat pumps is normally stated as a Co-efficient of Performance (CoP) of say 4 to 1 which is essentially 400%.

This might seem a little nonsensical but the figure is valid and reached because one unit of electricity is used to generate three units of heat energy, or to put it another way, three times the heat energy is produced by only one unit of electrical energy, which is where this figure of 400% - or CoP of 4:1 in this case - is derived.

Most of the energy comes from the external environment, and only a fraction (1/4) comes from electricity - or as the diagram below shows, 75% of the heat is from the atmosphere.

The CoP gives us the ratio of heat energy out to the electrical energy input as shown in the diagram below. In other words 4 units of heat in all are delivered to the property for one unit of electricity supplied.

75% of the heat produced is generated from the atmosphere

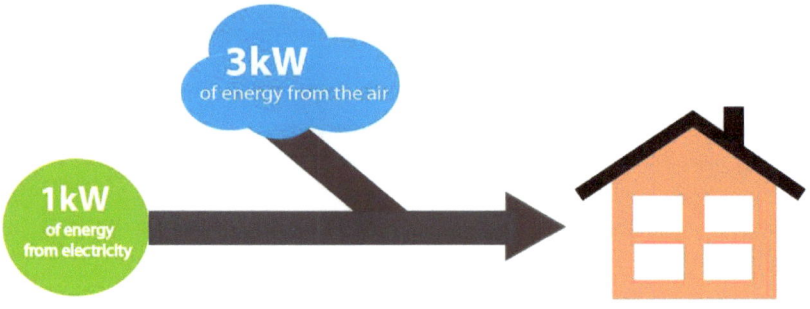

In a conventional electrical resistance heater, an immersion heater for example, all the heat is produced is from the input electrical energy so we could give this a CoP of 1:1 or 100%.

If we compare our heat pump (with an efficiency of say 400%) with Oil or Gas boilers which have efficiencies of perhaps 95% at best, it is easy to see why a heat pump is an attractive proposition.

If we also consider that we can produce some, or even all, of the heat pumps electricity requirements by say wind or solar power then the argument for using a heat pump becomes a strong one.

From an environmental perspective a heat pump also scores well.

At present around 6.5% of our (UK) energy comes from renewable sources and as this percentage is likely to increase our heat pump will become increasingly green.

As electricity is our dominant "couldn't live without it" energy source, we are likely to see a growth in the take up of heat pumps.

The amount of CO_2 produced (or saved) throughout the day does vary over the seasons and also each 24 hour period. If you want to see how much CO_2 is being produced at any one time check out www.realtimecarbon.org

Seasonal CoP or Seasonal Performance Factor (SPF)

Seasonal Performance Factor is similar to CoP, but is an average figure taken over the year and as such is perhaps more useful as a working figure.

It is usually lower than the stated CoP as the efficiency of an air source heat pump will drop when it's colder.

How many kilowatts is a 15 kilowatt heat pump?

The European standard for testing and rating heat pump performance is EN14511.

The main benchmark used to give the kilowatt output of air source heat pumps is A7 / W35 (for air source heat pumps)

The A7 bit refers to an ambient air temperature of 7° C whilst the W35 relates to a flow temperature of 35° C or put another way, your heat pump will produce 15 kilowatts when it's 7° C outside which will give a flow temperature of 35° C.

Remember our Air source heat pump extracts heat energy from the air so the ambient temperature directly affects the performance of our heat pump hence the SPF as noted above.

A benchmark is fine but you do need to check manufacturers' specifications rather than just rely on the product name which does not necessarily reflect the heat pumps output.

For example, referring to a couple of manufacturers data and part numbers, you might think a ***15 AS would be a 15 kilowatt heat pump but in fact at our A7/W35 standard it's rated as 7.3 kilowatts. Another manufacturer has a product name of LC-15 which (at A7/W35) produces 15 kilowatts.

There's no problem with these product names but is just something you need to be aware of.

Sizing a Heat Pump

As it happens, conveniently for UK residents, the bulk of our heating demand is around 7° C so the A7 bit is a reasonable basis for rating.

A heat pump works best with a low temperature distribution system - for example under floor heating which generally operates at around 35 to 40 degrees (which I suspect is where the W35 part of the kilowatt reference rating came from).

We can get much higher temperatures from our heat pump but at a gradually reducing efficiency so if we can work with a lower flow temperature then so much the better.

If we were sizing a heat pump to replace a boiler, and be used with radiators in an existing installation, it may be necessary to increase (over size) the radiator area in order to operate at the lower flow temperatures which the heat pump works best at.

Heat pumps are generally sized in one of three ways.

The first method is typically referred to as "Monovalent" where the heat pump is large enough to cater for all of the year round heating demands. This is probably the most energy and carbon efficient option, but requires a heat pump carefully sized for the job.

It is also best used in conjunction with under floor heating, LST's, Smart rads, or simply over-sized radiators as mentioned earlier.

Another option would be to use a heat pump which contains a backup electric (immersion type) element which kicks in to provide help to the heat pump when required.

The third option is a Bivalent system.

This combines a heat pump with for example a boiler or other heat source. This is often used where there is an existing heating system in good condition but the customer wants to reduce their CO_2 emissions, use of fossil fuels, and of course energy bills.

The system switches over to the boiler at a predetermined point (temperature) when the heat pump is unable to cope with the demand at that time on its own.

In a bivalent system, (or one with the backup immersion), the contribution of a heat pump to the total heating demand needs to be quantified.

A bi valent heating system

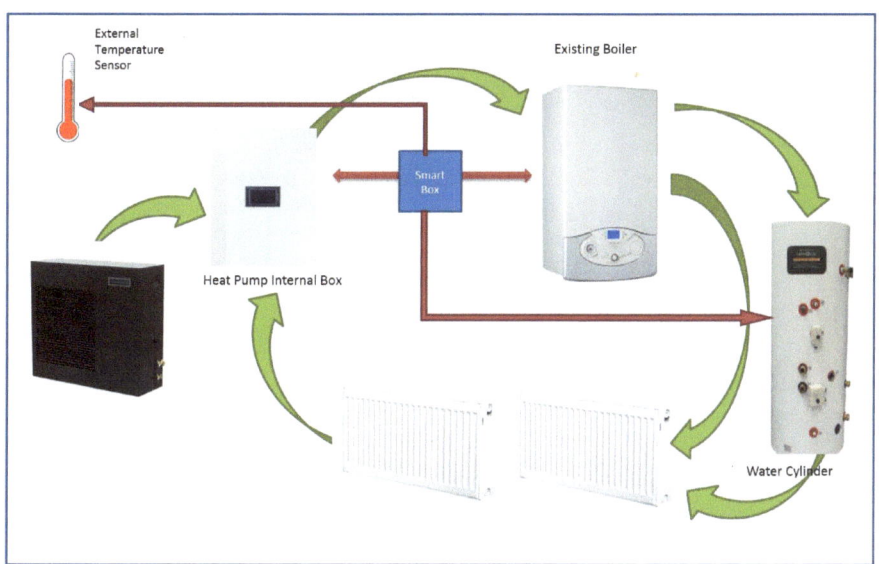

Picture kindly supplied by Lailey & Coates

Whichever system method is proposed you will need to rely on your suppliers' ability to size the heat pump system correctly.

Essentially they need to first calculate your buildings heat loss which is what it says - how much heat your property is losing through the roof, walls, floor, draughts etc.

A buildings insulation is key to its energy efficiency and the principle of insulation is to slow the path of heat through the buildings fabric - its walls, windows and so on.

Essentially, if we know how well a material conducts heat we can calculate how much heat we lose - the figure we need to calculate this heat loss is the "U value".

On a new build it's easy - the architect will have all of this information but on an existing building it's a little more difficult. However, there are representative typical "U" values for building materials, for example an external wall made of 9 inch thick solid brick has a U value of 2.2, whilst an insulated block and brick cavity wall has a U value of 0.6.

A uPVC double glazed window has a U value of 2.9, whilst a uPVC krypton filled triple glazed window has a U value of 0.8.

As you can see there is a lot of difference between common building materials and methods and this is only the tip of the iceberg. It would take a lengthy book to give this subject justice.

The other part of the heat loss calculation is draughts or technically speaking, the buildings ventilation loss. A draught means heat is being lost somewhere. You might have the best triple glazed windows in the world but if the frame isn't properly fitted (and of equally good quality to that of the glazing) then you are not getting the maximum benefit.

So before you spend money on a new heating system spend some time and money on keeping the heat in.

There are companies who can do things like thermal imaging and air tightness testing but a walk around your property on a windy day with an incense stick will tell you a story (and make your house smell nice). The usual suspects are window and door seals, service entry ducts, loft doors, sockets etc. Just a few pounds spent on some expanding foam and sealing mastic can make a world of difference.

RENEWABLE ENERGY - A BUYERS GUIDE

Thermal image of a property showing where heat is being lost

It is most important to do heat loss calculations to properly size the heat pump and your supplier should have the necessary experience and tools to do so.

Ultimately the heat loss calculation will give a figure of watts per square metre - something like say 50 watts per square metre. Therefore if your building had a heat loss of 50 watts per square metre and your property was 100 square metres you would need a 5 kilowatt (50 X 100) heat pump to cope with this heat loss at times, plus of course the hot water demand.

There are benchmarks for various types of buildings available which give a watts per square metre, (heating and electricity), and a number of software packages available to help with these calculations so there are plenty of tools at the suppliers disposal to carry out this task.

That isn't the whole story but hopefully will give you a flavour of what is involved and what your supplier should be doing.

 A reputable company will happily give you the calculated heat loss figure for your property but you should not necessarily expect this to be done free of charge.

Air Source Heat Pumps

These absorb heat from the outside air and extract heat in the same way our fridge does.

We know that the refrigerant used in heat pumps boils at very low temperatures which is why our Air Source Heat Pump (ASHP) can continue to work in sub-zero temperatures.

As shown in our heat pump process diagram we then take this extracted heat to warm our homes and give us hot water.

Typically a heat pump can produce temperatures of around 55 degrees for space heating and hot water.

We can use this heat to heat radiators, under floor heating systems and other types of emitter we talked about earlier. However, the higher the temperature the heat pump has to produce, the more energy it consumes so if we can heat our home using a lower temperature, the heat pump will be more efficient and use less energy. Heat pumps work most efficiently at low flow temperatures - this is true for all types of heat pump - and pretty much any other form of heating for that matter.

An Air Source Heat Pump from Lailey & Coates

Defrosting

Air source heat pumps have fans that draw air over the evaporator and extract heat. If you put your hand over the air outlet of a heat pump you will feel just how much colder the air that's had its "heat" extracted is. At low ambient temperatures the evaporator will likely be below freezing point and moisture in the air will freeze on its surface and form a coat of ice.

This is perfectly normal and of course the lower the temperature and the higher the humidity the more often this "frosting up" will occur. So, just like your freezer the heat pump needs to defrost itself.

Various sensors automate this defrost process so everything's taken care of.

One of the most common methods to deal with this process is called reverse cycle defrost. Essentially your heat pump runs in reverse taking heat out of your heating system to defrost itself quickly - where this defrost process takes it heat from is the subject of much debate in the heat pump world. *A heat pump which runs in reverse may also be able to provide cooling as well.*

For our reverse cycle defrost you could for example, take the heat out of say your radiators by running the heating system in reverse, and use this heat to defrost your heat pump. This is not uncommon but has a downside in that you are interrupting your heating and lowering the temperature of your building. However, the amount of time involved for defrosting, maybe only three or four minutes per hour, and the amount by which you lower your building temperature, may be so small as to be negligible - or not - it's all weather dependent.

The other common method *(which is the focus of the industry debate)* is to have a water cylinder fitted which the heat pumps heats - this water cylinder then supplies your central heating system. These cylinders are known as buffer tanks because they supply a buffer of hot water and even out temperature differences. In this configuration, when the heat pump needs to defrost, it takes heat out of the buffer tank rather than directly from your heating system.

There are as always pro's and con's to each approach - the main downside of having a buffer tank is the space needed to accommodate this water cylinder. That said, buffer tanks aren't usually very big, tending to start at around 50 litres of water upwards.

However, if your heat pump is also being used to provide domestic hot water as well as space heating (which most will be) then you will need a hot water cylinder anyway so whatever the arguments for and against buffer tanks, you'll need space for a cylinder in any event.

There are cylinders on the market which are combined buffer and hot water cylinders which will reduce the overall amount of space needed.

Inverter and non Inverter heat pumps

Although we discussed inverters earlier the inverter referred to here is not for the same purpose as the inverter for Solar Photovoltaic etc.

Another hotly debated subject in the heat pump world.

Essentially inverter driven heat pumps have what you can think of as a throttle, kind of like your cars accelerator which increases and decreases the amount of power your heat pump provides in line with your heating demand.

A non-inverter heat pump (sometimes also referred to as "DOL - direct on line" units), don't have this accelerator - they are simply on or off.

However, as with all things in life there are yet again pro's and con's to each system.

One of the main points to note is that DOL air source heat pumps get more efficient as the ambient temperature rises but inverters get less efficient. If you wanted to heat a pool the standard on / off system may well be the best choice.

The question of inverter / non inverter heat pumps may also determine whether your supplier advises the use of a buffer tank or not. The heart of a heat pump is a compressor and being a motor, constantly turning it on and off isn't particularly energy efficient.

A non-inverter heat pump really should have a buffer tank as the heat pump then only need respond to the buffer tanks requirement for heat, which is more of a constant than your home heating system may be. In other words, a buffer tank will reduce the amount of times your heat pump has to turn itself on or off.

An inverter driven heat pump can modulate (accelerate / decelerate) in line with demand so maybe doesn't need that buffer tank.

Your will have to rely on your suppliers experience to advise upon this aspect.

However, if in doubt, fit a buffer tank - or better still, a thermal store.

So we've chosen our Air Source Heat Pump, it now needs to be located somewhere.

They are usually (but not always) fitted externally, and as we mentioned earlier, they absorb heat from the outside air so it is preferable (but not vital) that they are positioned with a southerly aspect to benefit from maximum solar gain.

However, it's not critical and the ultimate position will probably be dictated by where any existing heating system is located or the general buildings layout.

That said, there may well be areas that we can scavenge heat from - for example, extractors from laundry rooms or kitchens. This is more likely to be of use in commercial applications but any opportunity to scavenge heat should be exploited. It is surprising just how much heat is simply ejected by a commercial kitchen *(unless we're already using a mechanical heat recovery and ventilation system which is a whole other story)*.

Your installer should be sympathetic to positioning with regards to neighbours or potential noise issues. Heat pumps **ARE NOT** particularly noisy, and in fact UK regulations state the maximum (very low) permissible noise levels, especially with regards to neighbouring properties.

In residential areas sound levels of less than 45dbA must not be exceeded - and to put that in context think of a dishwasher in operation - not really noisy but does emit some noise.

Therefore, even though todays heat pumps are quiet it's best not to stick them outside your bedroom window, especially if you are particularly sensitive to noise as some people are

A complete Air Source Heat Pump system incorporating Solar Thermal

Integrated home heating solution – heat pump with solar and under floor

1. Heat Pump
2. Indoor Control Unit
3. Hot Water Cylinder
4. Radiators
5. Underfloor Manifold
6. 3 Port Diverter Valve
7. 2 Port Zone Valve
8. Auto Air Vent
9. Solar Panel + Manifold Station

Ground Source Heat Pumps

Fortunately the principals of operation are pretty much the same as for air source heat pumps.

The key difference is where the heat pumps gets it's "heat" from.

The other obvious difference is that unlike air source heat pumps, ground source units are normally located internally so are afforded better protection from the environment.

Essentially, ground source heat pumps get their heat from pipes (ground arrays) which are buried in the ground, placed in water, or placed in boreholes.

As we know, the term "heat" is relative so if the ground source heat pumps collector piping is in water which is only at say 4 degrees, to a refrigerant, this is still "hot".

Water also has a high specific heat capacity which means a small amount of water contains a lot of heat.

Most of these ground arrays are closed loop - that is to say they contain a fluid which stays in the ground loop pipework and is circulated to collect "heat".

Horizontal ground loop

Slinky ground loop

The pictures above show two common types of ground array - horizontal loop and slinky.

The horizontal collector has the lowest capital cost but as you may realise, takes up a considerable area and involves a substantial amount of groundwork. As a rough guide it takes about 100 square metres to produce from 2 to 5 kilowatts so we are talking about a lot of pipe.

The main advantage of "slinkys" is that less ground area is required, but it is important that the slinkys don't extract too much heat from a small area such that the ground cannot recover its heat.

Slinkys can also be laid vertically (as shown in the next picture) in a narrow trench and there are specialist machines to enable the "slot trenches" to be dug very quickly.

The principle of ground collectors is that the ground temperature (2 metres down) is about 10 degrees year round which is fine for a heat pump.

This temperature is higher than an air source heat pump will have available during the winter months, which is where ground source heat pumps have scored in the past giving them a higher CoP than their air source counterparts.

It is important that the ground array is buried sufficiently deep to benefit from these "high" ground temperatures with 1.5 to 2 depth metres being the norm.

Slinky ground loop laid vertically

The alternative to horizontal ground arrays are boreholes which don't really take up any surface space and are typically 50 to 120 metres in depth.

A hole is drilled, the pipe/s are put into the hole which is then "back filled" with what is called grout - typically this is a mixture of silt, sand and bentonite. As this grout surrounds the pipework it has to be thermally efficient to transfer the maximum available heat so the grout material is critical to the design.

The only way to be sure of the type of ground that may be encountered by drilling, is to have a "trial hole" drilled first so costs may be incurred before the project even begins.

Drilling boreholes is very much a specialist job.

Just as with air source heat pumps, ground source units extract heat from their surroundings, in this case from the ground (or water). However, if the ground array pipes are put too close together then too much heat may be extracted and the ground may never recover and can freeze, making the ground array useless. This can be clearly seen as permafrost on some of the not so well designed ground loops.

The type of ground will also have a significant effect on how much heat we can extract, and, how quickly the ground can recover.

The size of the ground collector is also critical and as a rule the bigger the collector the better, but cost and available area will likely dictate the maximum collector that can be accommodated.

As we said right at the beginning, most renewable energy comes from solar energy either directly or indirectly. In this case solar energy heats the ground which gives us our energy source so the amount of "heat" will depend upon shading - heavy shading should be avoided if possible.

Rain is our friend as it helps move heat down into the ground so impermeable surfaces should be avoided. Wet ground is our friend as moving ground water helps to "recharge" the ground.

The last commonly used source of heat for ground source units is water. We are talking about a large volume of water and as a rough guide it takes about 33 metres of pipe per installed kilowatt, so a 10 kilowatt heat pump would need around 300 metres of pipe to provide its energy source.

A water source collector

The above picture shows one type of water source collector which as you will see is very similar to a "slinky".

The pipes are weighted down and the whole unit (this type is called a pond mat) is pulled out into the lake, or whatever water source it is, and let sink naturally to the bottom.

Water tends to be a good heat source as the water will tend to be naturally changed, (constantly recharging our heat battery), and also every square millimeter of pipe will be in contact with the water so the transfer of thermal energy will be good.

However, just as with ground arrays, care is needed to ensure we don't extract too much heat and freeze our water source and that the volume / depth of water is sufficient to cater for the heating demand

Ground Source Heat pump Schematic

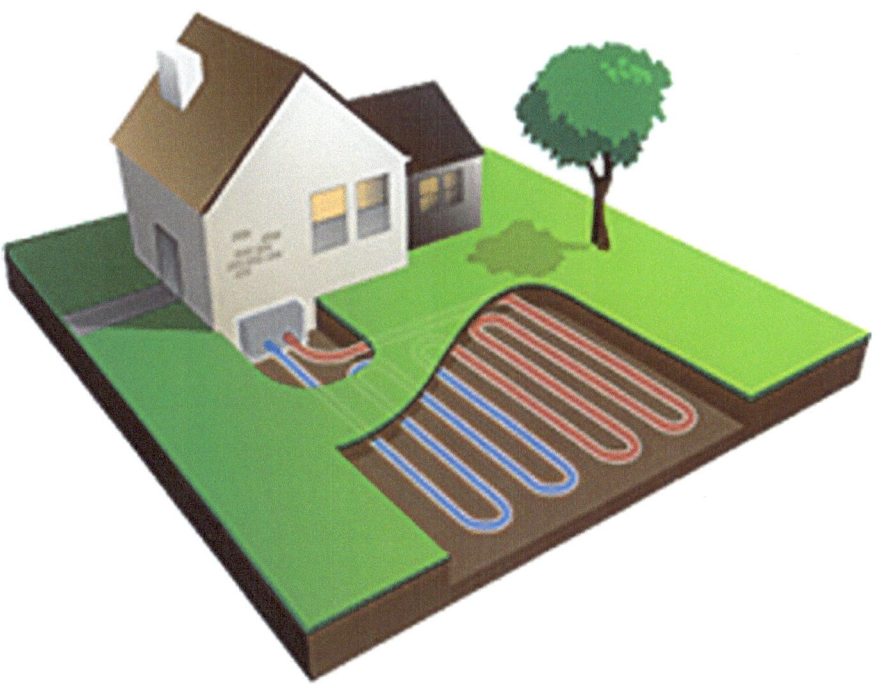

The main reason that ground source heat pumps have been historically preferred (from an efficiency perspective) to air source units in the past, is that ground units have tended to give higher CoP's benefiting from the higher source temperatures available in winter.

However, time moves on and new technology has narrowed this gap as have hybrid heat pumps which we'll discuss shortly.

Ground Source Heat Pumps in action

This lake is used to supply heat to The Salmon Centre, Laxey, Isle of Man.

The centre has a working model of the famous Laxey Wheel (Hydro in action), a shop, and is a popular function venue for both tourists and locals alike.

The centre is owned by Mr Stewart Clague, founder of SCS, who are the largest mechanical and electrical contractors on the Isle of Man.

The company embraces new technology and when Stewart decided to refurbish the Salmon Centre a ground source heat pump was chosen to heat and cool the building which SCS installed.

The ground source (or perhaps more accurately water source) collector consists of 1.25 kilometres of 32mm pipe and if you do the sums (using the previously noted figure of 33 watts per metre) you'll see we have approximately 36 kilowatts of low grade heat available which perfectly matches the heat pump and building requirements.

There are five separate "loops" of 250 metres, each loop being set onto a pond mat as noted earlier.

RENEWABLE ENERGY - A BUYERS GUIDE

The Ground Source Heat Pumps at The Salmon Centre, Laxey

Notice the grey pipes top and bottom which brings the water to and from the lake

The reason a heat pump with cooling facility has been chosen isn't because of the tropical climate on the Isle of Man. As the centre provides a venue for weddings, *(and other such somber occasions)*, when full of dancing bodies it can get surprisingly warm so some comfortable cooling is welcome.

By the time this book is published a 9 kilowatt hydro electric generator will have been installed making this building carbon neutral.

This is a great example of a natural resource, in this case water, not only supplying the heat source but also electricity.

Hybrid Heat Pumps

These products are a not new but only recently seem to have come to the attention of the UK market.

The principle of operation is similar to as an air source heat pump, but, the evaporator (which collects the heat) is a special type of solar collector which is filled with a refrigerant.

Just as with solar thermal systems the "solar collector" extracts heat from solar irradiation. However, the BIG difference here is that the solar collectors are filled with a refrigerant which has many more times capacity to absorb heat than conventional solar thermal panels.

Solar Thermodynamic collectors

Picture kindly supplied by Hyrax Solar

The picture above is a project carried out for the MOD and is new training centre built on one side of a WW2 aircraft hangar.

This system caters for the heating demand supplying an underfloor heating system.

Whilst these "solar collectors" benefit from solar gain they can work year round being able to extract heat from rain, wind or snow.

Because these special solar panels contain this refrigerant they will also work at night making them a strong contender in the heating and domestic hot water market.

I'm not sure whether they would be classed as a heat pump system, a solar system, or both, but combining these two technologies has resulted in high levels of efficiencies and our use of natural resources.

However, the "Solar Thermodynamic" name tag seems appropriate given that they combine solar collectors and heat pump technology.

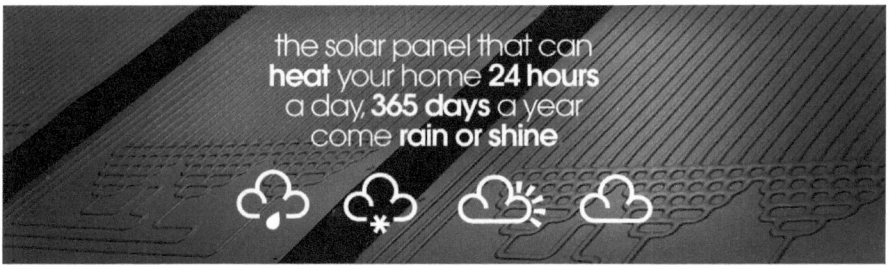

These solar panels are typically fixed to a roof or wall but I imagine they could be put anywhere that we could scavenge heat from.

The noted CoP's are very high and on a par with many ground source heat pumps but without the groundwork of course.

Whilst on the subject of combining technologies, as we have seen, producing your own electricity from solar photovoltaic or wind power is actually relatively easy.

This is one reason that I believe the take up of heat pumps will increase dramatically - if you can produce electricity you can potentially offset the small amount of electrical energy your heat pump is using.

It's pretty unlikely that you could produce your own oil or gas for a boiler.

Biomass - more than just wood

Winter evenings curled up in front of a wood burner is probably the image conjured up by romanticists when the subject of Biomass is mentioned but biomass is more than just wood.

It refers to anything that has been recently living or existing waste such as crops, rapeseed, grass, vegetable matter, sewage, hemp *(which has other popular uses)* and of course wood. However, as far as the domestic market goes then wood is king - well it would be wouldn't it .....

Wood is an obvious renewable source and is also considered carbon neutral. This is because the quantity of CO_2 emitted during combustion is equal to the CO_2 that was absorbed by the tree during its growth. In fact, burning wood only releases the same amount of CO_2 as it would have if left to decay naturally on a forest floor.

That said, it is easy to "conveniently" overlook emissions caused by the cultivation, manufacture and transportation involved so locally sourced product is important to retain a credible low carbon image.

As long as new trees (or crops) are planted to replace those that are used for fuel it is very sustainable but a long term strategy is vital.

The obligatory log burner aside, wood fuel comes in other guises such as wood pellets or chips.

Wood Pellets

Wood Chips

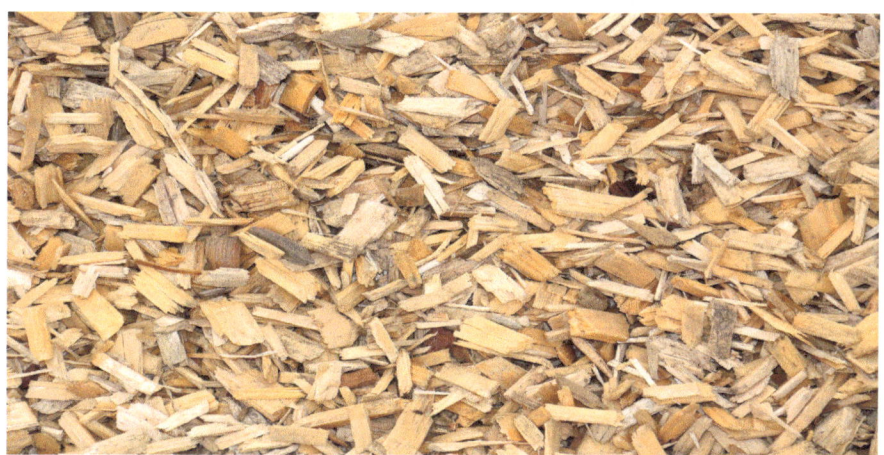

Wood fuels can be used in a modern central heating system whether it is logs, pellets or chips.

This type of heating system is very common in France where wood fuel is abundant and readily available

If you like the idea of a wood burner but are not so keen on humping logs into the house, there are wood pellet stoves available which are a close facsimile to log burners whilst offering the convenience of remote control operation from your armchair.

The cost of pellets is generally higher than logs as pellets are manufactured to a tolerance and standards for moisture content, size and calorific value exist. That said, a well-seasoned hardwood can be an expensive commodity.

The main factors that will affect cost is the method of delivery, "wood miles" involved, and the volumes you purchase. The greater the volume the cheaper the cost - but - remember you will need sufficient dry space to accommodate your wood fuel.

Whether a boiler, log burner or pellet stove, they will all require regular cleaning and ash emptying - the frequency of this will greatly depend upon the wood fuel you are using.

Just as with a boiler, annual maintenance is also a must

The chimney and flue pipes on a wood burning stove or boiler must be swept regularly and HETAS (the official UK body recognised by Government to approve biomass and solid fuel domestic heating appliances), recommend this should be done at least twice a year. Once before the heating season and once at the end of the heating season is good practice.

There a fair bit to consider before taking the plunge into biomass heating and every situation is different.

The commercial sector has taken to biomass quite strongly in the UK - in fact Drax Power Station in North Yorkshire is working towards becoming predominantly biomass fuelled.

This trend is likely to continue as biomass does give significant savings in carbon dioxide emissions, and buzzwords such as Corporate Social Responsibility will no doubt continue to "fuel" this market growth.

A wood pellet boiler from Effecta - Sweden

Voltage Optimisation

Not a renewable but currently fashionable as an energy saving device so we could put it under the heading of carbon reduction.

Some may remember in the good old days, (way back in the 90's), when electrical products were rated at 240 volts.

Older electrical equipment made for continental Europe was rated at 220 volts.

However, we got "harmonised" with the EU in 1995, after which all products made for the European market had to operate satisfactorily at 230 volts plus or minus 10% which is between 207 and 253 volts.

This didn't however mean that the electricity suppliers had to turn down the voltage coming into your property, and the reality of the situation is that incoming supplies are usually 240 volts plus.

All of your electrical products will operate perfectly well at this voltage but critically the most efficient supply voltage for these appliances is 220 volts which is what a Voltage Optimiser does - it reduces the incoming voltage to 220 volts.

For the "techies", if we look at this in electrical terms, Ohms law states $P = V \times I$ which means power = volts multiplied by amps.

So, take a light bulb and put it into 2 equations with different voltages and the same current (amps) and we get:

240 (volts) x 1 (amp) = 240 watts

OR

220 (volts) x 1 (amp) = 220 watts

Referring to the earlier section "What's a watt", remember you are charged in kilowatt hours for the electricity you consume and the above example shows clearly there is a reduced wattage which equals reduced running costs.

However, in this case it also means that the amount of light output is reduced which may not necessarily be a problem (in fact it may increase the lifespan of our light bulb) but it should be noted.

In this case electricity would be saved but at the cost of a slightly reduced light output.

In a similar vein, if we reduce the voltage to an electric kettle it will take longer to boil which is not an issue, but in this case we are using less power but over a longer period so we won't save any electricity. These types of electrical products are known as resistive loads.

It is products such as fridges, freezers, washing machines, dishwashers etc where the greatest savings will be made. This is because they are known as inductive loads.

Some voltage optimisers have a fixed voltage adjustment whilst others regulate the supply voltage automatically but the end result is pretty much the same.

Voltage optimisers are simply fitted between your electrical meter and your consumer unit (AKA distribution board AKA fuse board).

It's a relatively simple concept and fit and forget technology that is reasonably affordable.

As far as the electricity savings go every property will differ - manufacturers generally state that savings will be somewhere between 5 and 15%.

A voltage optimiser being fitted

Summary

I hope you have enjoyed reading this book and for those who are thinking of buying into renewables, that you have gained an insight into the various technologies on offer.

There is no doubt that these technologies will advance, and new ones emerge, however, one things is for sure - they will save you money and reduce your carbon footprint.

It's a fair bet that energy prices will continue to rise so your return on investment should get better with time.

In the not too distant future I hope to follow up with a revised and revamped edition once both of my typing fingers have healed sufficiently. Until then - thank you for listening.

Marc Asker.

About the author

Marc lives in the UK with his wife Patricia and their rescued German Shepherd dog Brockley (who has more friends on Facebook than Marc does).

As ardent Dog lovers, after more than thirty years of marriage Marc is used to playing second fiddle to their continuous canine family.

Marc has been involved with renewables for more than ten years - designing systems, installing, consulting, specifying and as an accredited instructor, has taught renewable technologies both in the UK and France.